WITHDRAWN

D0875870

FIRE SAFETY

Smoke Alarms

by Lucia Raatma

Consultant:
Roy Marshall
State Fire Marshal of Iowa

Bridgestone Books
an imprint of Capstone Press
Mankato, Minnesota

Bridgestone Books are published by Capstone Press
818 North Willow Street, Mankato, Minnesota 56001
http://www.capstone-press.com

Printed in the United States of America.

Library of Congress Cataloging-in-Publication Data

Smoke alarms/by Lucia Raatma.

p. cm.—(Fire safety)

Includes bibliographical references and index.

Summary: Discusses the importance of having smoke alarms in the home, the different kinds of alarms and how they work, and what to do when they sound.

ISBN 0-7368-0196-0

1. Fire detectors—Juvenile literature. [1. Fire detectors. 2. Fires. 3. Safety.] I. Title. II. Series: Raatma, Lucia. Fire safety.

TH9271.R33 1999

628.9'225—dc21

98-44798
CIP
AC

Editorial Credits

Rebecca Glaser, editor; Timothy Halldin, cover designer and illustrator; Kimberly Danger, photo researcher

Photo Credits

David F. Clobes, cover, 6, 8, 10, 12, 16, 18
David Clobes Stock Photography/Pat Thom, 14
Gregg R. Andersen, 20
Unicorn Stock Photos/Kathi Corder, 4

Table of Contents

The Danger of Fire 5
Loud Alarms 7
Where to Put Smoke Alarms 9
Parts of a Smoke Alarm 11
Smoke Alarm Batteries 13
Testing Smoke Alarms 15
Keeping Smoke Alarms Clean 17
When a Smoke Alarm Sounds 19
Smoke Alarms Save Lives 21
Hands On: Test Your Smoke Alarms 22
Words to Know 23
Read More 24
Internet Sites 24
Index 24

The Danger of Fire

Smoke and flames from fire can hurt you. You must leave your home quickly if a fire starts. A fire can start and spread quickly. Smoke alarms warn you if a fire starts. The warning tells you to get out quickly.

smoke alarm

a machine that senses smoke and warns people by making a loud sound

Loud Alarms

A smoke alarm senses smoke. Smoke alarms make a loud noise when smoke is in your home. This noise can wake you if you are asleep. Ask an adult to test the smoke alarms in your home. Then you will know what the noise sounds like.

Where to Put Smoke Alarms

Most homes need more than one smoke alarm. Your home should have a smoke alarm on each level. A smoke alarm should be on the ceiling near bedrooms. All smoke alarms should be on ceilings or high on walls. Smoke rises to the ceiling when fire burns.

vents
horn
sensor
battery

Parts of a Smoke Alarm

A smoke alarm has four main parts. A battery or electricity powers a smoke alarm. Air and smoke enter a smoke alarm through holes called vents. A sensor inside an alarm senses smoke. The horn makes a loud noise if the sensor detects smoke.

electricity

power that comes into your house through wires; electricity powers some smoke alarms.

Smoke Alarm Batteries

Many smoke alarms use batteries for power. Remind adults to change the batteries twice a year. Adults should remove batteries only if they put in new ones. Smoke alarms without power will not sound an alarm.

ST ME
TEST M

Testing Smoke Alarms

Adults should test smoke alarms once each month. Most smoke alarms have a test button. The alarm should go off when an adult presses this button. Remind an adult to test the smoke alarms in your home.

Keeping Smoke Alarms Clean

Dirty smoke alarms might not work. Dust, grease, or bugs can block vents. Smoke alarms with blocked vents cannot sense smoke. Remind an adult to clean the smoke alarms in your home once each year.

Warren
Two Worlds One TRIBE

When a Smoke Alarm Sounds

Leave the house quickly when a smoke alarm sounds. Your home might be smoky if it is on fire. Stay low under the smoke. Follow your family's escape route. Meet your family outside.

escape route

a planned way to leave your home during an emergency

Smoke Alarms Save Lives

Smoke alarms sometimes go off when there is not a fire. But always pay attention to smoke alarms. Smoke alarms can save lives. The sound of an alarm gives an early warning so you can get out quickly.

Hands On: Test Your Smoke Alarms

Smoke alarms cannot sense smoke if they do not work. You can help an adult test smoke alarms each month.

<u>What You Need</u>

Smoke alarms in your home
An aerosol testing spray for smoke alarms
An adult

<u>What You Do</u>

1. Press the tester button on the smoke alarm. Or spray the alarm with the aerosol testing spray.
2. If the alarm sounds, you are done. Go to the next smoke alarm in your home. Repeat step 1.
3. If the alarm does not sound, make sure the power is connected or the battery is working.
4. Replace the battery if necessary.
5. Check the vents for dust, grease, or bugs.
6. Use a vacuum cleaner to clean the smoke alarm if it is dirty.
7. Test the smoke alarm again.
8. Smoke alarms can break and wear out. If a smoke alarm doesn't work, replace it with a new one. Test the new alarm to make sure it works.

Words to Know

battery (BAT-uh-ree)—a small container that provides electricity; some smoke alarms need batteries for power.

electricity (ee-lek-TRISS-uh-tee)—power that comes to your house through wires; electricity powers some smoke alarms.

escape route (ess-KAPE ROOT)—a planned way to leave your home; escape routes should have two ways to leave each room and a place outside to meet your family.

smoke alarm (SMOHK uh-LARM)—a machine that senses smoke and warns people by making a loud sound

Read More

Loewen, Nancy. *Fire Safety.* Plymouth, Minn.: Child's World, 1997.

Raatma, Lucia. *Home Fire Drills.* Fire Safety. Mankato, Minn.: Bridgestone Books, 1999.

Internet Sites

Inside a Smoke Detector
http://www.howstuffworks.com/inside-smoke.htm

Sparky's Home Page
http://www.sparky.org/

United States Fire Administration (USFA) Kids Homepage
http://www.usfa.fema.gov/kids/

Index

battery, 11, 13
ceiling, 9
electricity, 11
escape route, 19
fire, 5, 9, 19, 21
horn, 11
level, 9
sensor, 11
smoke, 5, 7, 9, 11, 17, 19
test, 7, 15
vents, 11, 17
warning, 5, 21